A Home

What is a home?

It is a place to sleep in the dark, dark night.

EMC 4094

What is a home?

It is a place to keep dry in the wet, wet rain.

It is a place to keep warm in the cold, cold snow.

EMC 4094

What is a home?

It is a place to be safe from hungry, hungry hunters.

What is a home?

It is a place to raise a new, new family.

EMC 4094

What is your home?

Your home is a place where your family lives.

It's the place where you eat and sleep.

It's the place where you keep warm and dry.

It's a place where your family has fun together.

Draw your home here.

Do animals have homes?

Most animals don't have homes.

They don't need a special place to stay.

They go around looking for food.

Color the animal.

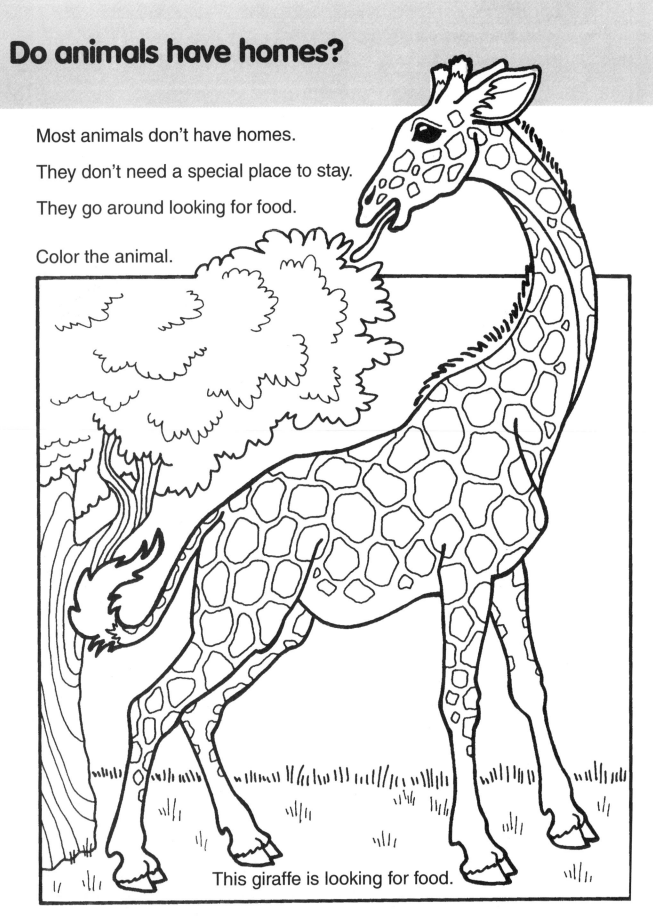

This giraffe is looking for food.

EMC 4094

Where do animals sleep?

Most animals sleep wherever they are.

They don't need a special place to stay.

Color the animal.

The walrus is sleeping on a rock.

Where do animals have their babies?

Most animals have their babies wherever they are.

Color the animals.

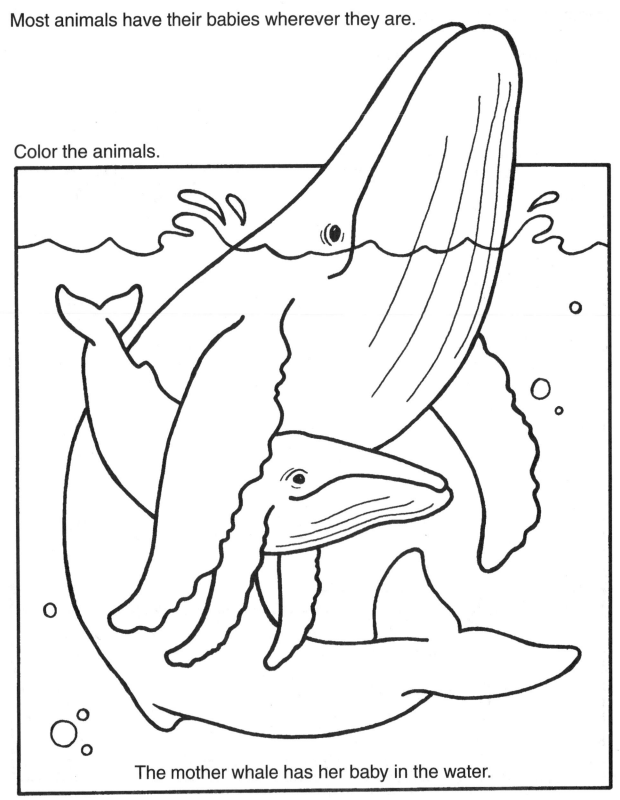

The mother whale has her baby in the water.

EMC 4094

Some animals do build homes.

They build homes to be safe.

They build homes to raise their babies.

Color the animals.

This bird builds a nest for her babies.

Animals build with many things.

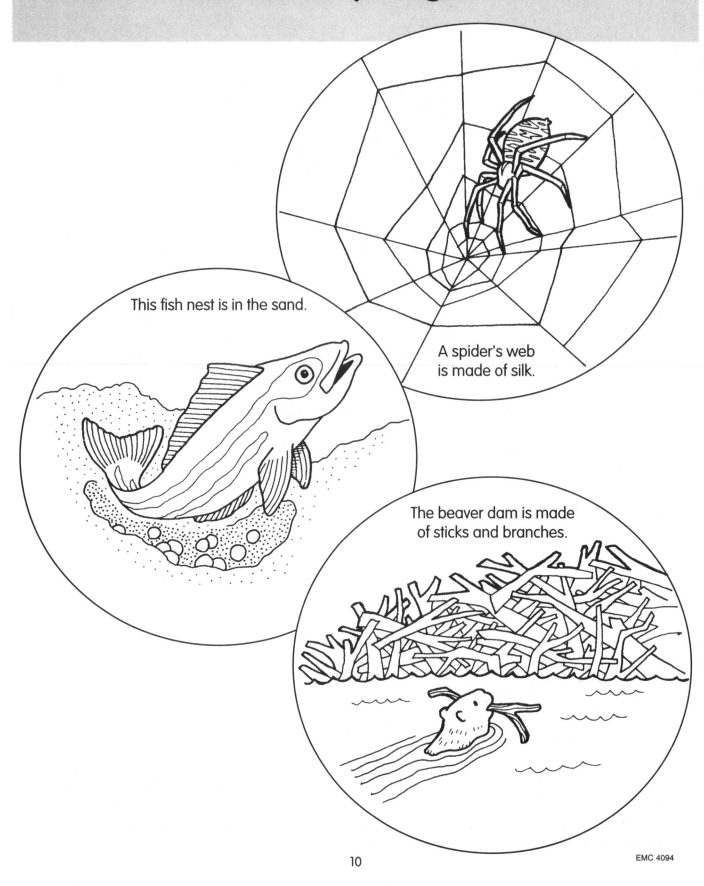

A spider's web is made of silk.

This fish nest is in the sand.

The beaver dam is made of sticks and branches.

EMC 4094

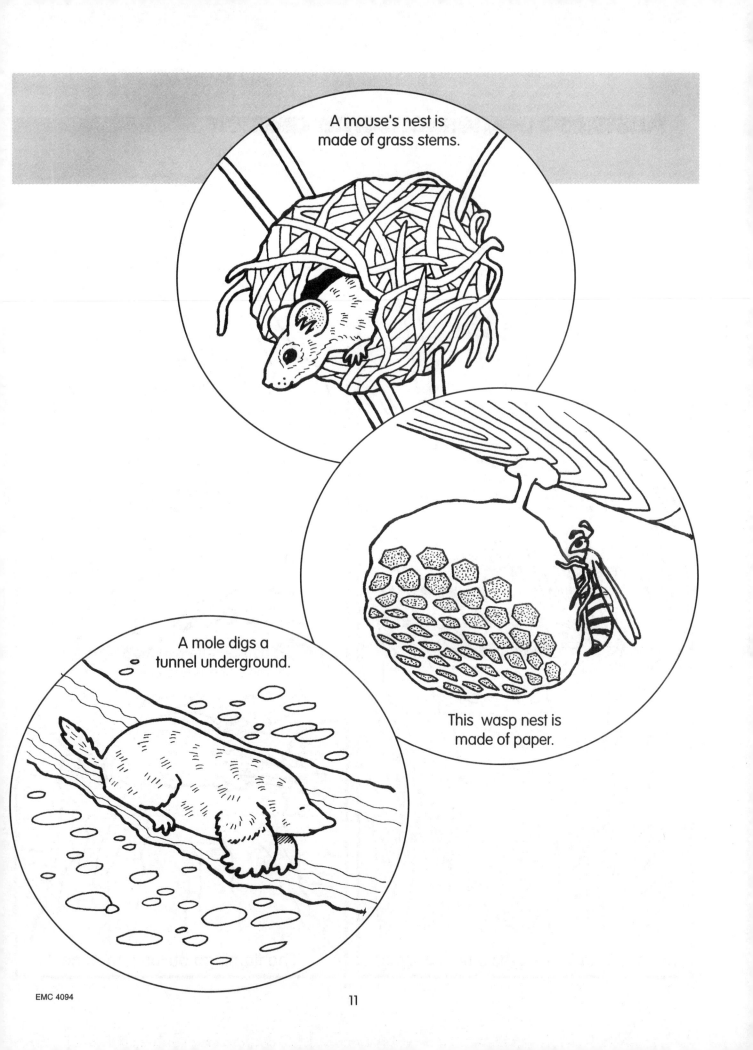

A mouse's nest is made of grass stems.

This wasp nest is made of paper.

A mole digs a tunnel underground.

Nests are homes for some animals.

Some birds build nests.

Nests are made of different materials.

The reed warbler weaves a nest of grass.

The flamingo builds a mud nest.

EMC 4094

Other animals build nests too.

A stickleback fish has an underwater nest.

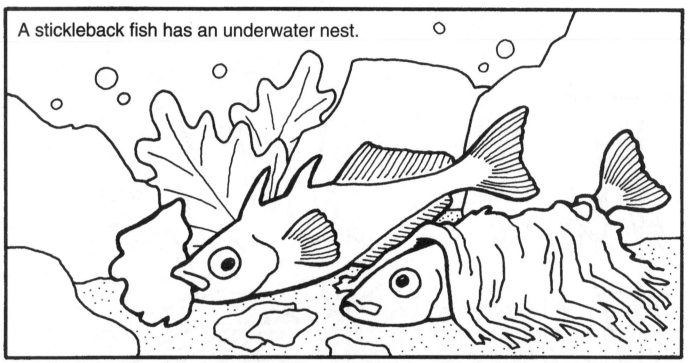

The alligator builds a nest of plants and mud.

Even some dinosaurs made nests long ago.

This dinoasaur built a round nest for its eggs.

EMC 4094

Find my home.

Match.

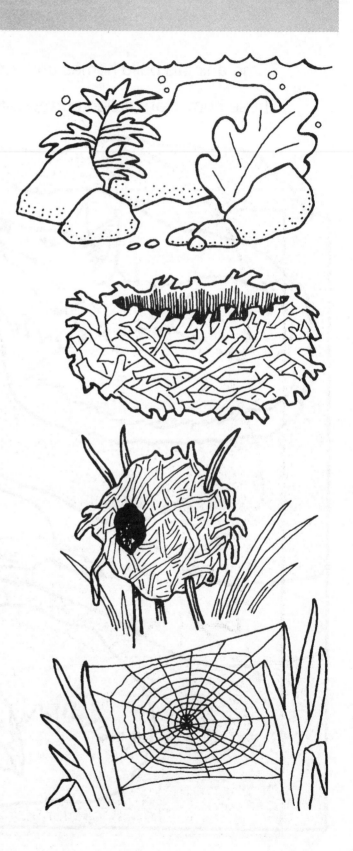

Some animals build underground homes.

Some animals build homes under the ground.

These homes are called burrows, tunnels, or dens.

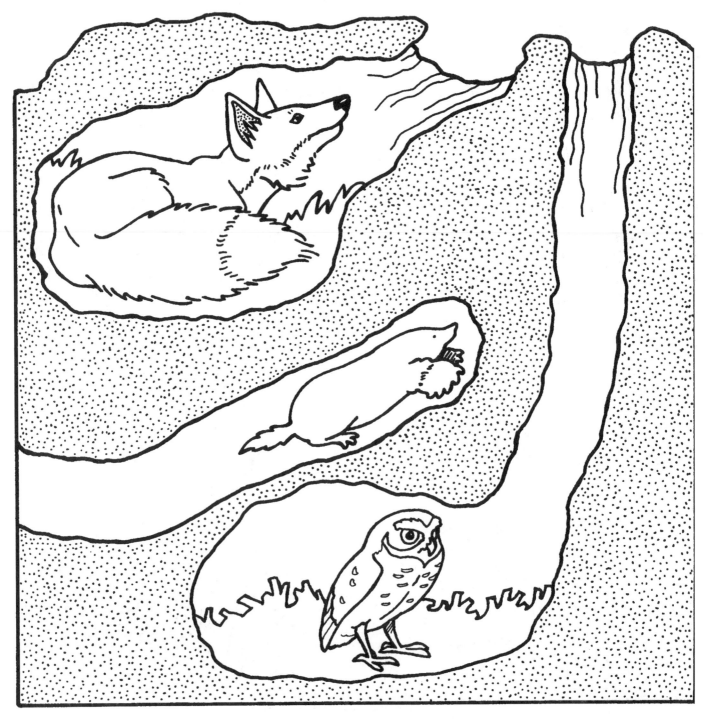

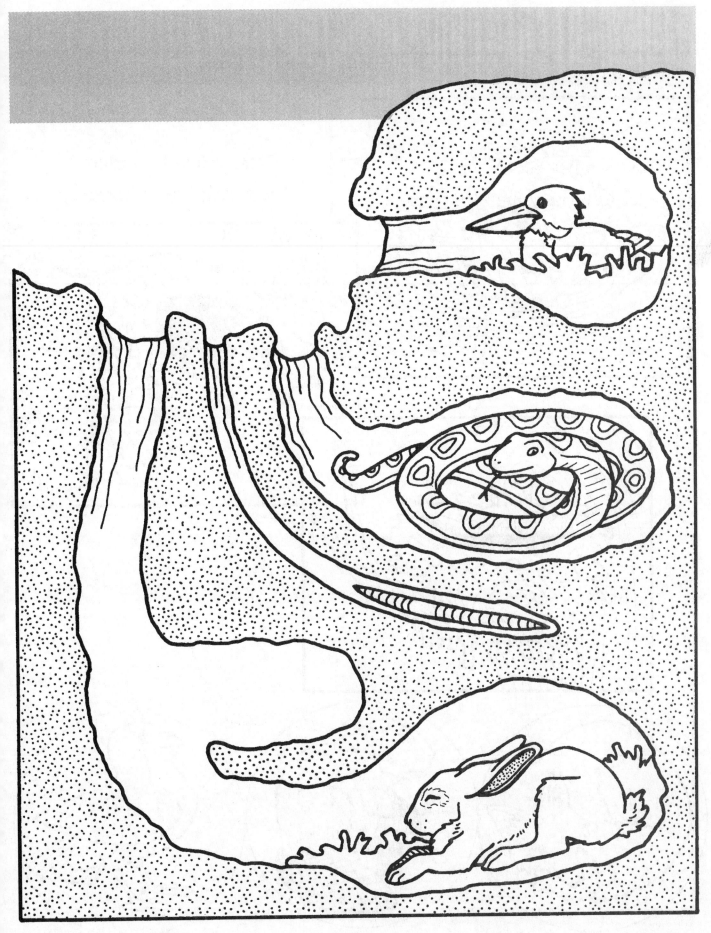

Homes are many shapes and sizes.

Draw a line to match the animal to its home.

18

Some animals live together in groups.

These animals live together.

They build a home together.

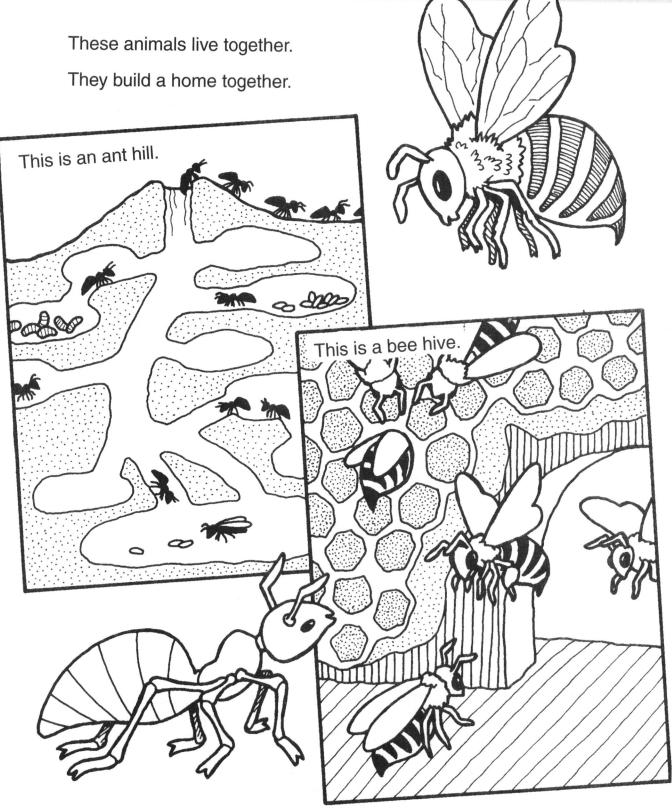

This is an ant hill.

This is a bee hive.

This is a prairie dog town.

Prairie dogs live together in groups too.

They dig underground tunnels.

How many prairie dogs do you see?

Color the prairie dogs brown.

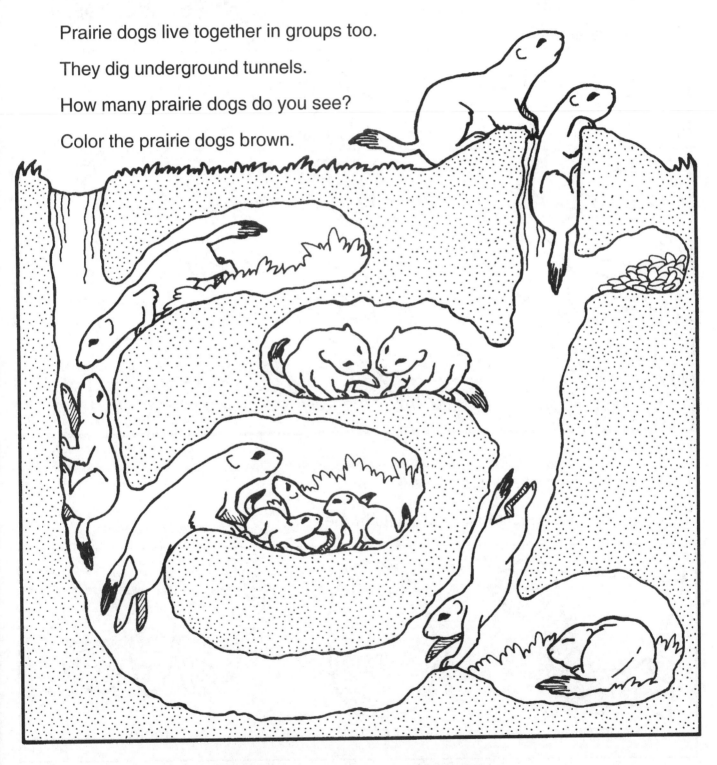

I found _____ prairie dogs.

Who builds a home with silk?

orb spider web

Some caterpillars make a home of silk.

The caterpillar changes into a moth in its home.

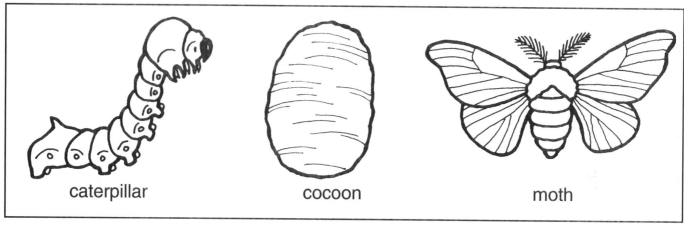

caterpillar　　　　　　cocoon　　　　　　moth

Circle the homes of silk.

　　　　　　　　　　　　EMC 4094

A platypus builds her nest in a burrow.

Caves can be homes too.

Color the bear.

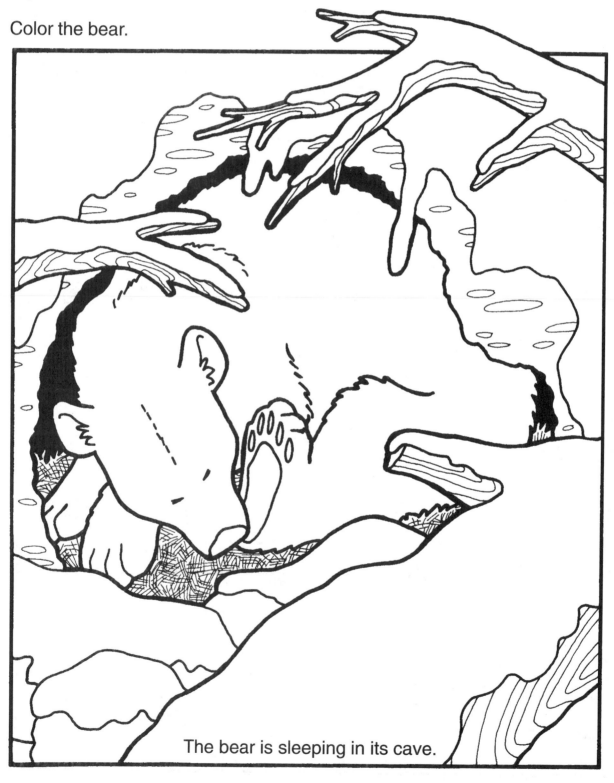

The bear is sleeping in its cave.

EMC 4094

Can you find the beavers?

How many beavers did you find?

Some animals borrow homes.

Parents: Help your child read this page.

Color the cactus green.

Color the owl brown.

Color the shell yellow.

Color the crab brown.

Elf owls move into empty
woodpecker holes in cactus.

Hermit crabs move
into empty snail shells.

EMC 4094

Some homes move from place to place.

Some animals carry their homes with them.

This turtle and snail have shells that are a part of their bodies.

They take their shells with them wherever they go.

They can pull into their shells to be safe from danger.

We can take a home with us too.

Put an X on the ones you have stayed lived in.

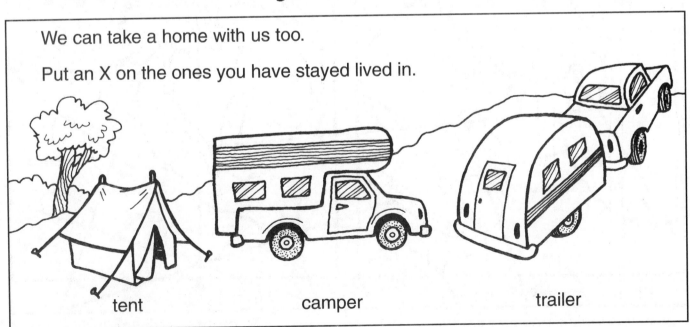

tent   camper   trailer

Can you find the animal in his home?

Color the • red.

Color the ▢ brown.

Color the ● blue.

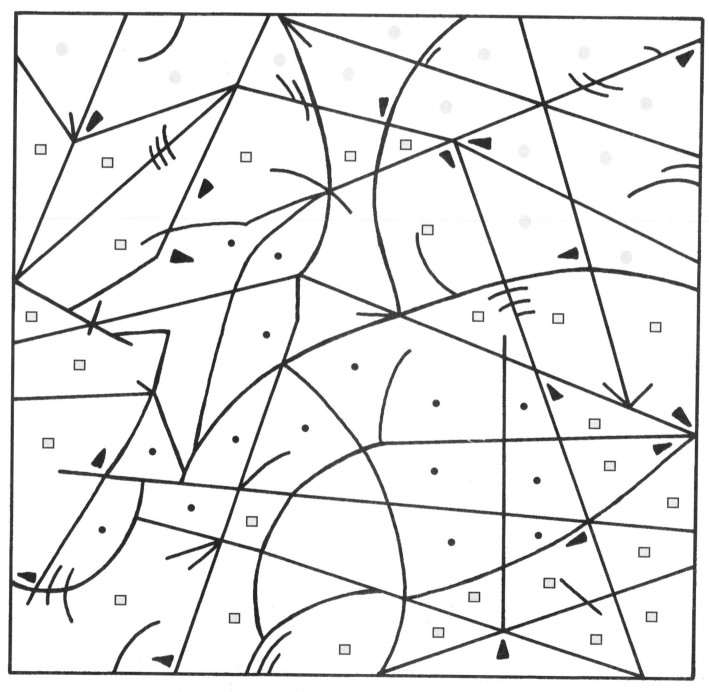

EMC 4094

Farmers make homes for animals.

Put the farm animals in their homes.

chicken coop

rabbit hutch

barn

pig sty

Our pets have homes too.

Parents: Explain to your child that he/she is to cut out the pictures on page 31 and to paste the pets in their homes.

EMC 4094

Parents: Your child is to use these pictures on pages 29 and 30.

Color  **Cut** 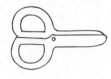 **Paste**

EMC 4094